COCONUT OIL
Diet and Obesity

Calixto López

COCONUT OIL

DIET AND OBESITY

**Calixto López
(2018)**

AUTHOR'S FOREWORD

There is a lot of talk about the functional properties of coconut oil, including its possible use to lose weight and combat obesity, but what is true about the latter, what evidence supports it? This is precisely what this book is about, finding answers to these questions, but with scientific evidence drawn from the publications of researchers and specialists who have delved into this complex and rugged field.

A key, a kind of code can guide us in our purposes, something related to its particular and original lipid profile, perhaps the best kept secret of coconut oil, the one that can help us or predict its conduct: MCTs and ketone bodies. With these tools we will focus on the essence of the problem, regardless of the result.

The book itself consists of three chapters, plus an introductory one. The first, in terms of content, deals with the original composition of coconut oil, the second on its role in the ketogenic diet as genesis of the research developed so far and the role of MCTs in it, and the last on its possible effects on overweight and obesity. In that order they are written, but if the reader wishes to vary it or omit some, we only recommend that before submerging in the last one he has some notion about the nature and composition of this oil and the metabolic role of the MCTs.

Editor's Note:

In the translation of the book into the English language, in addition to human labour, translation programs were used. We are sorry if you find any grammatical errors in the text.

CHAPTER I

INTRODUCTION

Vegetable oils are essential staple foods for the proper functioning of the human body. However, consumed arbitrarily, without knowing their lipid profile and nutritional properties, they can be harmful to health, which drastically reverses their possible role.

Each vegetable oil has a different composition or lipid profile than the others, which is the cause of its physical and chemical properties, and the possible applications derived from them. Of the components that form part of the oils, the triglycerides or esters of glycerine with fatty acids of different structure and nature stand out, which are the ones that define the particular properties of the oil.

Within the vegetable oils, the coconut oil presents very particular properties which have motivated that in the last times special attention is dedicated to it, in him very specific factors are conjugated related to his original lipid profile, in which saturated fatty acids predominate of average chain, very different to the one that exists in the common oils, in which they abound more the fatty acids of chain superior to 14 atoms of carbon.

Coconut oil is not a product frequently found in supermarkets and retail outlets like other common oils, although more than 2.8 MTM are produced annually in the world, preferably in countries with a tropical climate, and most of them with an emerging or developing economy. At the same time, it has taken interest in the media and scientific dissemination, including those related to the network, attending to a group of pharmacological properties attributed to it, including its possible effect on brain diseases such as epilepsy and Alzheimer's, for which due to their complexity there are no effective treatments.

In spite of the above, due to its lipid profile rich in saturated fatty acids, although of medium chain, it cannot be induced that coconut oil is not associated with cardiovascular diseases, although defenders and detractors appear in this sense. Also, considering that medium chain fatty acids are easier to metabolize in the body, there are those who argue that the energy associated with them is consumed quickly, avoiding or reducing their storage in adipose tissue, and therefore playing a positive role in the fight against obesity and overweight, combined with a proper diet.

We studied an oil with a particular lipid profile in which the prevailing fatty acids are those of medium chain: **lauric** (C12:0), **capric** (C10:0) and **caprylic** (C8:0), which endows this product with an endless number of interesting and original properties.

Today there is much evidence to consider coconut oil as a functional food or product, even though some sectors of the scientific community doubt its efficacy, or in the prevention or treatment of certain conditions, and in the technology sector opinions are divided between the supporters or beneficiaries of this industry and the harsh or brutal competition.

This is influenced by different factors. On the one hand, the popularity that this oil is having in the propaganda means as an effective cure in the treatment of multiple ailments, without sufficient conclusive evidence to endorse it, so that in some cases tends to exaggerate, or attribute properties and health benefits that this oil does not have, or its effect is less than desired, so that in the same bag are included others that in effect does. In this sense news circulate, even videos, about hundreds of conditions that can be treated with this "wonderful product", which more than a defence contributes to its questioning.

Other sources, even specialized, tend to look at coconut oil, or demonstrate, that it does not influence, or does so negatively in the treatment of various conditions, or that it can constitute a risk factor, as in cardiovascular diseases due to its high content of saturated fatty acids. In the latter case, opinions are very divided, with several international institutions such as the American Heart Association (AHA) showing a critical role, independently of the fact that no conclusive evidence has yet been found on their role and possible effect on them.

On the other hand, it is common in the conduct of people that if they have a condition and talk about the wonders of a product, try to find in it the remedy for their ills, as well: if you tend to baldness its use for hair, if there are epidermal conditions, in the treatment for the skin, if you are obese to lose weight, and so on, with which you go to the product assessing in advance results that have not proven, which may cause you to appreciate qualities that are not, or create predisposition not to find what you expected.

And it is that this product or universal panacea for the treatment of all evils does not exist, although there are drugs, functional foods or other products of wide spectrum given its complex chemical composition that in effect can be of more general use than others, as it happens for example, with *allium sativum L.* (garlic) that for containing organosulphorate compounds in its composition, plays different functions as hypolipemiant agent, hypoglycemic, antioxidant and bacteriological, etc.

With oils, the problem takes on a very particular character, as it is forgotten in their values that they are complex mixtures of variable composition, in which numerous substances are present, although we simplify their denomination as mixtures of triglycerides, although they are accompanied by other components with a generally beneficial effect on health, although not in all cases.

In this way, the secondary components that accompany olive oil and other vegetable oils complement their nutritional effect on the organism, and it should not be forgotten that although they are minor constituents, their action may be of relative intensity, or at least representative. Thus, on the negative side, peanut allergens, some of which may accompany the oil, although in small proportion, are capable of causing drastic effects in people prone to allergy to this food. A similar case occurred with rapeseed, which in relation to erucic acid considered a toxin, forced important research in the 60's of the last century to obtain a variety of plant with low amounts of this long chain acid, that is, canola, whose oil is currently one of the highest production and consumption in the world, over sunflower.

At the same time, tocopherols, polyphenolic compounds, vitamins, minerals, etc. that accompany virgin vegetable oils are beneficial for health, some of them with a marked anti-inflammatory action such as "oleocantal", recently isolated from some lots of Greek olive oil, with action similar to ibuprofen and which can be an effective form of use to help compensate the state of people who need this anti-inflammatory, even if their amounts in olive oil are too small to replace the treatment of the drug, or that is not found in all virgin olive oils.

With regard to coconut oil, it is found more than in scientific hands on many roofs: in the media, in public opinion, even in religious fanaticism, and in the minds of many simple people in need of effective remedies for their affections, who see in it a hopeful element in the treatment of their ills.

It seems to us, then, that the scientific community and specialists in the subject must take sides in this controversy and rather than fall into Byzantine lobbying and discussions, establish on scientific bases what may or may not be certain in the basic attributions that are made of this oil.

In a recent book we published on coconut oil, as part of a systematic study we are carrying out on each of the basic vegetable oils consumed in the world, we defended that coconut oil, more than anything else, is a vegetable oil with a particular chemical composition, which determines its basic properties and characteristics and as such had to be seen and used in what was really most useful according to that composition. We do not overlook in this study, its main characteristic related to its original lipid profile, composed of triglycerides of medium chain fatty acids (MCT) which is what gives it its physicochemical properties and its striking solid state at room temperature, as well as the effects on the organism derived from this profile.

With this vegetable oil, and with all those previously studied, we have tried to make prevail the idea of achieving the identity of each one of them, and that they are not seen as secondary products subordinated to one remedy or another, and in this case the coconut oil is the most accurate to be treated as such, bearing in mind that its production is ranked 10th. in the world, and is consumed by millions of people in the world.

This time the challenge is much more delicate, because the content may or may not agree with the opinions or assumptions of people from different social communities and even scientific, so as on other occasions, we will only rely on our judgement when there is evidence or conclusive evidence, or that result in axioms that do not require verification, and in this sense we will venture to try to demonstrate, if the evidence so demonstrates, that **coconut oil is a functional food** and can be useful in the prevention or attenuation of certain conditions, even if it does not reach the magnitude that is sometimes expected, or desired.

In particular, we will focus on the effect of coconut oil on overweight and obesity, taking as evidence the results of experimental research on animals and humans carried out by different specialists and published in articles in specialized journals, without in any of the cases being able to make generalizations as broad as desired, but if they serve as a basis for the treatment of the issue in question.

As a preamble to the focus on the effect of coconut oil on obesity and overweight, we had to deal with two important issues that serve as a basis for understanding the nature and properties of this oil, which are those that determine its action on the body in terms of its composition or lipid profile, and the background that underpins its use from research conducted on the ketogenic diet, and ketone bodies, and the role played in it by the MCTs, **the best kept secret of coconut and its oil.**

CHAPTER II

COMPOSITION OF COCONUT OIL

1. Coconut Oil

Coconut oil is a very peculiar vegetable fat, which differs greatly from other basic oils in its lipid profile, which confers properties and characteristics somewhat special, especially if for that reason, its attention is currently focused on the controversy of whether or not its use in food is beneficial or not, and its effect on human health.

The controversy, as we expressed in this case, is served, and some value and overvalue it as a panacea with multiple benefits for the welfare and metabolism of the body, due to its rare and high composition of saturated fatty acids of medium chain (MCFAs), much higher than that of other common oils. Others, by virtue of its high concentration of saturated fats, even higher than that of African palm oil, consider it to be a risk factor for cardiovascular disease (CVD), and above all for maintaining adequate cholesterol levels.

The common oils: **sunflower, canola, palm, soy, maize, peanut, cotton and olive**, base their properties on a main central axis related to the concentration of the basic prototypes of fatty acids: **palmitic, stearic, oleic**, the first two saturated (SFAs) and the third monounsaturated (MUFAs). In some cases, the significant presence of polyunsaturated fatty acids (PUFAs), such as **linoleic** acid, with two double bonds in the hydrocarbon and **linolenic** chain, with three, also stands out. But in any case we are referring to fatty acids with equal hydrocarbon chains, or greater than 16 carbon atoms.

Coconut oil, however, presents a particular lipid profile in which saturated fatty acids with a medium hydrocarbon chain prevail, where **lauric** acid (C12:0) with a concentration of 47% or more stands out, and others with a shorter chain length: **caprylic** (C8:0): 8%, **capric** (C10:0): 6%; to which **myristic** acid is added (C14:0): 18%, which give this oil very special properties and characteristics, in addition to the fact that this high indicator of saturated fatty acids, over 90%, affects its physical properties, especially the relatively high melting temperature 24-26 C, which makes it present as a white solid substance in countries with a cold or temperate climate, but not in southern countries, or in countries with a warm climate, in which it can be presented as a slightly pale yellow or colourless liquid.

This composition of coconut oil is not the only element that determines that we make a differentiated assessment of it, because different compositions show other oils such as peanut oil with relatively significant values of arachic acids (C20:0); 1.5%) and behenic (C22:0), 3.0%) or that of the original rapeseed that still contains a certain proportion of erucic acid (C22:1), or soybean with levels close to 50% linoleic acid (C18:2), and other indicators that give it texture, taste, and characterize these oils.

So if it was only the problem of composition that attaches importance to coconut oil, it might not show relevance and any analysis of it would be done focusing mainly on its lipid profile. There is another extremely important factor and it is not even the economic one: it constitutes the media attention that is being given in the media, including of course the net, and by different authors, especially by those who magnify their beneficial properties for health or others who vehemently reject this assumption. What consciously or unconsciously can cause problems, especially in people more likely to believe blindly in what is heard or written.

In essence, before referring to the polemic, we can define coconut oil as a vegetable fat obtained from the white mass of the coconut, the fruit of the *Coconuts Lucifera Linn*. Extracted by pressing (**virgin**) and then purified, bleached, deodorized and in short, refined (**refined coconut oil**).

Refined coconut oil (**RBD**) is presented as a pale

yellow or colourless liquid at temperatures above its melting point: 24-26C, or semi-solid with a lard-like texture at temperatures slightly below the melting point, even hard and brittle at temperatures below 15C. Virgin oil preserves the smells, taste and aromas of the fruit, while refining tends to be odourless and tasteless. The price in the European retail market exceeds that of common vegetable oils, even olive oil.

1.1- Basis of Controversy.

The nature and composition of coconut oil, where MSFAs predominate, is considered by some to be better assimilated by the body and therefore easier to metabolise, and that there is not enough evidence to consider that its saturated character may be associated with cardiovascular disease or cholesterol-elevating blood and other indicators related to atherosclerotic damage, such as low-density lipoproteins (LDL). Although in recent articles on human experiments, reference is made to the fact that it does raise these two indicators, but also high-density lipoproteins (HDL), thereby counteracting their possible negative effect. But no conclusive evidence has yet been drawn from all this.

His theory is also supported that lauric acid and caprylic acid are found as part of breast milk in proportions slightly higher than 6 and 2% respectively, which are also found, although to a lesser extent in cow's milk. In addition, and associated with breast milk, it has certain antimicrobial properties that could help the body's defences. It is true that this effect has been reported on some types of micro-organisms, but in humans more evidence would be needed to establish reliable correlations.

2. Lipid Profile of Coconut Oil

Until now, in this introductory part, different fatty acids have been frequently mentioned as components of coconut oil, so it is advisable to focus for the moment on their lipid profile:

Composition of fatty acids in coconut oil (g/100 g oil)

C8:0 Caprylic   8
C10:0 Capric   6
C12:0 Lauric   47
C14:0 Myristic 18
C16:0 Palmitic   9
C18:0 Stearic   2.5
Total SFAs   90.5

C18:1 Oleic   7
Total MUFAs   7

C18:2 Linoleic 2.5
Total PUFAs: 2.5

It is necessary to emphasize that in the market we are presented with several types of coconut oil that differ slightly in their composition. Preferably we will stop at the virgin coconut oil obtained by cold pressure on the ground coconut (copra) mass, which contains the original basic ingredients of the coconut mass without being subjected to heating or refining processes, and the RBD oil, which is refined coconut oil subjected to purification processes similar to those of other refined common oils, although with some nuances.

2.1 Coconut oil RBD.

The **RBD** coconut oil is presented as a very light yellow liquid, at temperatures above 26 C, below this temperature is solid, white, odourless and free of strange aromas and flavours.

Physically it has a melting temperature of 24C, or slightly higher, depending on its composition, origin and refining methods, but never higher than 27C.

From the chemical point of view it must present the following characteristics, in agreement with the established international norm:

1. Iodine value $g(I_2)/100$ g = 8.0 – 12,0
2. Maximum (lauric) acid number: 0,06
3. Meq peroxide value O_2/kg maximum: 10,0

4. Composition of fatty acids %.

Caprylic (8:0): 6.0 - 10.0
Capric (10:0): 5.0 - 8.0
Lauric (12:0): 44.0 - 50.0
Miristic (14:0): 16.0 - 20.0
Palmitic (16:0): 8.0 - 11.0
Stearic (18:0): 2.0 - 4.0
Oleic (18:1): 4.0 - 11.0
Linoleic (18:2) 1.0 - 3.0

As soon as you look at the lipid profile of coconut oil, you can see the high proportion of fatty acids with a smaller molecular chain size than those found in other basic oils, in essence, of the most common oils: sunflower, rapeseed, African palm, soy and maize.

These acids are: (C12:0), lauric (47%), and (C14:0), myristic (18%), also that the concentrations of acids (C8:0) caprylic (8%) and (C10:0), capric (6%), are not negligible, and finally that the concentration of (C18:1) oleic acid is much lower than in any of the other edible oils, which in the least of the cases is always above 15%.

For example, for the following fats, the concentrations of oleic acid are around these proportions:

Lard: 35-40 %.

Butter: 22 %.

Soybean oil: 20-25 %.

Corn Oil: 25-30 %.

Sunflower oil: 30 %.

Palm oil: 38 %.

Tallow: 40 %.

Rapeseed oil: 45 %.

Olive oil: 65-70 %.

Normally, in vegetable oils, the composition of the main fatty acids with a chain of less than 16 carbon atoms is relatively insignificant, which indicates that we are dealing with a very particular lipid profile.

However, similar compositions of fatty acids such as coconut oil are found in other palms as can be seen in the following table, which compares the acid profiles in % of palms: *Oleosa, Babasu and Coconut*.

Fatty acids   P. Oleosa   P. of Babasu   P. of Coconut *

Fatty acids	P. Oleosa	P. of Babasu	P. of Coconut *
Caprylic	6	4,5	8
Capric	4	7	6
Lauric	47	45	47
Myristic	16	16	18
Palmitic	8	7	9
Stearic	2,5	4	2,5
Total SFAs	**83,5**	**83,5**	**90,5**
Oleic	14	14	7
Total MUFAs	**14**	**14**	**7**
Linoleic	2,5	2,5	2,5
Total PUFAs	**2,5**	**2,5**	**2,5**

*Source: Belitz and Grosch (1997).

Also the oil extracted from the palm kernel (central coprose part of the fruit of the African palm), has a composition similar to that of coconut oil, so we see that in the main fatty acids of medium chain, this oil has the following profile:

2.2.-Lipid profile of RBD palm kernel oil (%)

Caprylic (8:0): 1.9-6.2
Capric (10:0): 2,6-5,0
Lauric (12:0): 40,0 - 55,0
Miristic (14:0): 14.0 - 18.0
Palmitic (16:0): 6,5-10,3
Stearic (18:0): 1,3-3,0
Oleic (18:1): 12,0-21,0
Linoleic (18:2) 1,3 - 3,5

The similarity is remarkable, although the latter oil has a higher proportion of oleic acid, also reported among its trade indicators the presence of very small amounts of other fatty acids: caproic: (C6:0), linolenic: (C18:3), among others. However, a comparison between both palms would not be advisable, given that in the technology of obtaining palm oils is used all the fruit, not just the central coprose seed, although these data refer to the oil of this almond.

The fact that the proportion of a monounsaturated fatty acid such as oleic acid in coconut oil is relatively low and lower than in other oils could mean that it provides less protection for CVS, and because the acids with the highest representation in coconut oil are saturated, they could also have a negative atherogenic effect on LDL and cholesterol concentrations.

For all these reasons it would be surprising if coconut oil had any representation in the edible oils market, and rather its use was aimed at the cosmetics industry, where there seems to be evidence of its benefits for the treatment of skin and hair.

On the other hand, and in a practical sense, it would seem more beneficial from the economic point of view the use of oil in the food industry, given the benefits of it, especially in the manufacture of jams and other related products, taking into account its relatively high melting temperature and smoke point, as well as that of its hydrogenated butter.

However, in relation to these aspects, and independently of what was discussed at the beginning about its effect on health, the reality is that this oil is produced and marketed in the world on a significant scale, with a volume of 2.8 MTM, in the 2016-2017 campaign (10th worldwide) slightly lower than olive oils (3.0) and corn (3.7). It is therefore necessary that we take into account a series of parameters that we will study later and that justify this apparently anomalous fact, but first let us pause for a moment in the characteristics of the plant and the goodness of the coconut tree.

3. Coconut palm

The coconut tree, whose botanical nomenclature is *Cocos lucifera Linn* is a type of palm of the family *arecaceae*, reaches a height of about 30 m and produces a fruit of great size: the coconut. It is considered a native of Asia, regardless of some controversies about its possible origin in America. The main producers are: Indonesia, Philippines and India among many others, as it is a tropical plant that has spread widely throughout the planet.

The fruit, given its high resistance, is an incessant sea voyager responsible for the main vegetation of many islets and atolls in the Pacific, where it has been carried by typhoons, storms and sea currents. Once it touches land, even if it is sandy, it germinates and is capable of clinging with its roots to the arid slippery terrain.

In these poor soil fertility conditions, the coconut tree grows to the slenderness and height that makes it a beautiful plant, sometimes solitary, but which characterizes the typical tropical landscapes of the islands.

Its wood is water-resistant enough that its logs have been used to build piers for small boats, and also as wood obtained laboriously by the locals, which becomes the main planks that cover the doors and walls of their houses.

The leaves of the coconut tree are large, and can measure up to more than 3 m long. They can be used in roofs and walls of rustic houses: ranches, bohios, huts, etc.

The coconut, fruit of great size, can present in several colourings: green and yellowish and of varied size. It needs relatively high temperature and humidity, conditions that occur in tropical regions, although it is capable of growing under certain conditions in areas with subtropical climates. It is a very resistant tree, able to face strong winds, but it does not support the cold, or the height. The one that accepts high salinities allows him to compete with success with other plants and that appears in beaches and sandy lands.

The coconut produce a fresh water with a pleasant and characteristic taste, which in recent years has been bottled, which facilitates its marketing in different parts of the world. The white mass (copra), inside the coconut, grows during maturity to reach hardness and consistency, and contains between 60-70% fat.

The yield of the coconut per unit of cultivated area is much lower than that of the African palm and according to data from crops in the Philippines, is of the order of 5TM per hectare.

Composition of the Coconut Fruit

In the coconut as fruit we can differentiate the following components.

Shell: 15 %.
Fibre: 43%.
Copra: 30%.
Water: 12%.

In coconut copra:

Oil: 65%.
Pulp: 17.5%.
Water: 17.5%.

The fruit of the coconut tree can weigh up to 2kg and within it several parts can be highlighted:

Thick, hard shell (exocarp)
Mesocarp: fibrous part
Endocarp: brown part containing the pulp
Endosperm: white mass that hardens as the fruit ripens.

The main product is the mass (copra), although bottled coconut water constantly expands its production and demand.

The coconut is marketed as fresh fruit, when it is approximately 6 months old, when its water content yields between 250 and 500 ml. In order for the dough to reach an adequate weight and thickness, it is necessary to wait more than a year or for the fruit to fall to the ground on its own, in other words, when it dries.

The dough is used for different purposes, not only to produce oil, but can be eaten directly or as is common, grated, how it is used in confectionery, pastry and in general for jams. A very original and showy ice cream is obtained from coconut contained in its shell, free of the fibrous outer shell (endocarp), this is called coconut glaze. One way to ingest coconut water is at intermediate stages of maturity, when it reaches greater sweetness and has formed a layer of soft white mass of good flavour. Coconut water is an isotonic drink.

Without any further processing, a milky liquid is extracted from copra, which when crushed and squeezed is of great nutritional value and can be used directly as a pure or mixed drink, as well as for culinary purposes.

In addition to fat (60-70%), copra contains fibre, generally soluble (10-11%), carbohydrates (3-5%), vitamins (E: 0.7 mg, C: 2.0 mg), and minerals such as K, Mg, P and Ca.

Also, although on a smaller scale, perhaps low for a fruit, contains carbohydrates and protein.

The cover that contains copra is used in industry to obtain high quality activated carbon, so that sometimes occurs, that the obtaining of oil and other coconut products are considered as by-products in obtaining this valuable adsorbent material, given the excellence of it and its high demand in the market.

As with other oilseeds, the pressed cake obtained as a

remnant in the extraction and refining of oil is used as animal feed, mainly for cattle.

Also the fibrous cover of the coconut can be used as fuel.

The coconut industry in the main producing countries of Asia has also left its negative remnant in the deforestation of extensive forested areas demolished to dedicate them to the intensive cultivation of this palm tree, although with less incidence than the African palm, but that it is necessary to bear in mind due to its damage to the environment and the climatic deterioration of the planet.

Like African palm oil, the high concentration of saturated fats in coconut oil slows its deterioration, especially rancidity, so it can be more than six months at room temperature without suffering appreciable oxidation, and much longer time under cooling, allowing its use in the production of ice cream, jams, etc..

However, the fact that coconut oil has a lower melting temperature than African palm oil means that for use in confectionery, and in the flour industry in general, it is subjected to catalytic hydrogenation for use, especially in regions with hot climates, which translates into a kind of margarine or coconut butter, with a melting temperature higher than 35C but with the handicap that trans fats are formed, with a negative impact on LCA.

5. Coconut Oil Extraction Methods.

Two basic methods are used: dry and wet.

Dry:

The dough is dried in various ways: by heating, by fire, sunlight or ovens, taking into account that in many places are used rudimentary and traditional methods of production. Then the copra is crushed and once obtained, it is pressed or dissolved, forming a kind of purée with high content of fibre and protein that is of low quality for human consumption, but not for animals, preferably ruminants. Sometimes this paste is unduly called coconut butter and is also a commercial item, but it is not exactly coconut butter or coconut oil, as it contains large amounts of fibre, moisture remnants and other components of the fruit. A portion of copra oil is lost in the extraction process.

Humid:

It uses crude coconut to create an emulsion between protein, oil, and water. Subsequently, the emulsion must be broken - a somewhat complicated aspect - to separate the oil. It can be done by prolonged heating, but the resulting oil is of very low quality and production costs are raised by having to increase the temperature during the process.

Modern centrifuges and cold pre-treatment with acids, salts, etc. are used.

In comparison, despite technological improvements, the wet treatment is less efficient and the yield is lower by more than 10%. On the other hand, technological equipment is more complex and expensive.

Also the method of use has to do with the process of maturation and harvesting of the coconut and its degree of dryness, it is always advisable to work with copra as mature as possible.

As in the extraction of other vegetable oils, n-hexane turns out to be a convenient solvent. After being treated, the mass is refined to remove free fatty acids and other substances that remain as impurities, and that can accelerate rancidity.

In essence, there are different processes of extraction and production, from the simplest, elementary and rudimentary, to others with advanced technology and equipment, similar to those used in obtaining most edible oils.

To obtain 100 L of coconut oil you need about a ton of crude coconut, or approximately 240 kg of dried copra.

A more modern RBD technique (refining, bleaching and deodorising) uses dry low hot-pressed copra, which extracts almost all of the oil (about 60% oil by weight of coconut) producing a crude oil not yet ready for consumption, so it must be refined with additional heating to remove polar substances of low molecular mass and then filtered. Refined oil is called RBD oil and is the most common oil on the market.

There are other techniques that include enzymatic processes, with which high quality oils are obtained. It is necessary to emphasize that refined coconut oil loses the flavour and smell of natural coconut, but does not suffer significant affectation in its lipid components, and if necessary are the MUFAs, according to what is recommended in some uses, it is not necessary to go to virgin coconut oil, for which there is not yet an appropriate certification, although in some countries, such as Germany, is working in this direction.

Virgin and "Extra Virgin" Coconut Oil

Although there are no guarantees to affirm that a coconut oil is extracted from the fresh fruit of the coconut tree, in the market oils are sold under the nomenclature of virgin and "extra virgin", which in essence are those that have not been subjected to refining processes and have been obtained only by pressing, grinding and separation by filtering, and according to information from producers, come from fresh coconut mass of freshly collected fruits.

The term "extra virgin" is inappropriate, since it is only authorised to name high quality olive oils, under organoleptic test by specialised tasters. Rather this may be due to a commercial ploy to obtain higher sales, as however this is incorrect and partly reprehensible. The term "ecological" is also questioned because in coconut plantations pesticides and herbicides are generally not used, due to the height of the plant, which is at the same time perennial.

In this situation the most correct thing to do is to refer to virgin coconut oils that have been obtained by wet or dry means. The rest corresponds to commercial techniques.

6. Hydrolysis of Coconut Oil.

Many coincide in affirming that the saturated fatty acids of medium chain components of the coconut oil, and that appear in this in the form of triglycerides, are easier to metabolize by the organism, even that they can diminish the indicators of obesity, including the levels of fat storage, since these are metabolized more quickly than those of long chain, and do not tend to accumulate in the adipocytes. In view of this and other industrial factors, studies have been carried out to isolate and obtain free form acids, not as triglycerides, which implies its hydrolysis according to the following reaction:

TAG + H_2O = AGL + Glycerine

This reaction in the organism is accelerated by means of enzymatic catalysis in which different enzymes intervene, but in the laboratory this can be modelled by means of the use of micro organisms that produce these biochemical catalysts, for example: *Candida cylindracea*, a process that takes more than two days and in which yields are obtained between 80-90%, corresponding to the concentration of fatty acids obtained similar to that of oil itself in acids of the same nature.

Final consideration

Finally, we would like to state once again that coconut oil is first and foremost a food vegetable oil, rather than a drug or an industrial product, so its consumption should be similar to that of any vegetable oil, that is, in moderate, not excessive quantities and in accordance with the demands and needs of the organism. For the expectations that are arising in the treatment of various diseases must be very attentive and make a safe use when there is evidence or conclusive evidence of its effectiveness against a particular pathology, meanwhile, recognize that so far is this: a vegetable oil rich in hydrocarbons saturated medium chain.

MEDIUM CHAIN FATTY ACIDS COCONUT COMPOUNDS

C8:0 Caprylic 8 %

C10:0 Capric 6%

C12:0 Lauric 47 %

C14:0 Myristic 18 %

Caprylic acid (C8:0). Octanoic

$CH_3 (CH_2)_6 COOH$

It is a liquid, saturated fatty acid with a medium hydrocarbon chain, made up of eight carbon atoms, including that of the carboxyl group. It is present with an approximate content of 7% in the oil of the African palm nut, and 8% in the coconut. It is also present in the milk fat of some mammals. Some physical properties are shown below:

M: 144.21 g/mol
Density: 0.91 g/cm^3
Melting temperature: 17,9 C
Boiling temperature: 237 C
pKa: 4,89

Caprylic acid has an antimicrobial action against certain pathogenic microorganisms such as: *Streptomyces agalactiae, S. dysgalactiae, Staph. aureus,* and *E. coli*, among others.

In acidic medium, at an acid pH of 4.8, it is used in specialized laboratories as a precipitant of a large number of plasma proteins.

Capric acid (C10:0). Decanoic

CH$_3$ (CH$_2$)$_8$ COOH

It is a saturated fatty acid of medium hydrocarbon chain length, constituted by ten carbon atoms, including that of the functional group carboxyl.

It is presented as a white crystalline solid with an intense smell, at room temperature, and melts at slightly higher temperatures.

Molecular Mass: 172.26 g/mol.
Melting temperature 31.6 C
Boiling temperature: 269 C
Density: 0.89 g/cm^3

The name capric acid derives from Latin, and refers to its smell similar to that of goats, in whose tissues it is found in a certain proportion, although to a greater extent in coconut oil as triglyceride, but in it its smell is not predominant, because if it were; this oil and the coconut mass in particular, would limit its use in cosmetics and pastry, etc..

Together with caproic acid (C6:0) and caprylic acid (C8:0) they make up about 15 % of the fat in goat's milk.

Although capric acid can be obtained by acid hydrolysis of fats, it is produced preferably by oxidation of decanol, an aliphatic alcohol of ten carbon atoms of chain length, by powerful inorganic oxidants such as chromium trioxide.

It is also of interest in the food industry as an antifoaming and for other purposes.

Lauric acid (C12:0). n-dodecanoic.

$CH_3(CH_2)_{10}COOH$.

Molecular mass: 200.32 g/mol
Melting temperatures: 42,2C
T. Decomp. 298 C
Density: 0.88 g/cm^3

It is a fatty acid saturated with a medium hydrocarbon chain, made up of twelve carbon atoms, including that of the carboxyl functional group. It is solid at room temperature but with a low melting point. It has a certain smell of soap, and in fact excellent hard and very foaming soaps are obtained from it due to its marked surfactant action, which is also one of its main uses, so it easily dissolves fats and non-polar liquids. It is also blamed for antimicrobial action.

It is found in a certain proportion in human milk fat (6.2%), ruminants such as cow fat (2.9%) and goat fat (3.2%).

Although it is present in the oil of several palms, it is in the coconut where it has acquired notoriety for being in it in a proportion close to 50%.

Together with myristic acid, they make up around 70% of the fatty acids in coconut oil, which is why it is considered, an oil rich in medium chain fats. As in some investigations they have been associated with the increase of low density lipoproteins (LDL) and therefore with atherogenic damage, is that some discrepancies of the positive effect of this on health, however, in recent times it has been related as an alternative product to attenuate or reverse Alzheimer's to a certain degree, although there is not enough evidence to the effect.

Like the acids mentioned above, lauric acid should not be ingested in its pure state, and in this case produces a strong irritation in the digestive tract.

Myristic acid (C14:0). Tetradecanoic.

$CH_3 (CH_2)_{12} COOH.$

Although, due to its chain length, myristic acid should not be included among the medium chain fatty acids, being at the border or dividing zone of both groups and being its uncommon presence in other vegetable oils, it is worth mentioning its main characteristics.

Myristic acid is a saturated fatty acid, solid at room temperature, with a medium to long hydrocarbon chain, made up of 14 carbon atoms, including that of the functional group carboxyl. It is not very soluble in water, but it is soluble in solvents of lower polarity.

Molecular mass: 228,4 g/mol
Density: 0.8622 g/cm³
Melting temperature: 54,4 C
Solubility: 1,07 mg/L

Its name comes from the nutmeg (*Myristica fragrans*), whose solid fat contains high amounts of this fatty acid (75%) in the form of triglyceride or trimyristine, as it is commonly called.

Its concentration close to 20% in coconut oil is considered a risk factor in cardiovascular diseases because of its positive correlation with low-density lipoproteins that transport cholesterol.

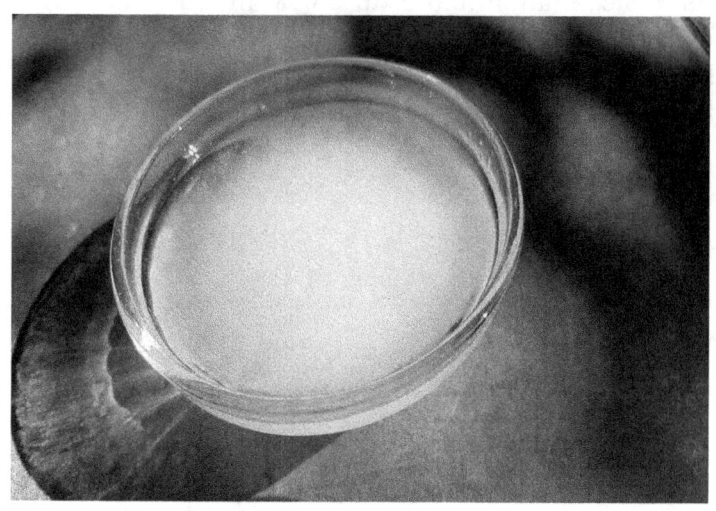

CHAPTER III

COCONUT OIL AND KETOGENIC DIET

Although the nature and purpose of the ketogenic diet arose outside of coconut oil, it is precisely because of it that we must begin our approach to the problem, since that is where studies and scientific research on medium chain triglycerides (MCTs) began, and their effect on different metabolic functions of the body, including cell combustion in areas as important as the brain. These medium chain triglycerides are found in a high proportion (more than 60%) as components of coconut oil, which gives transcendental importance to this product as a possible functional food, as well as the derivatives that can be obtained from it.

MCTs are the best-kept secret of coconut oil and the almonds of other similar palms, and they provide it with the surprising properties that it has in the treatment of various affections in very varied fields, especially those related to the metabolism of this type of lipids and their possible effect on the organism.

The ketogenic diet, whose name is related to the ketonic bodies or substances that are created during the metabolism of food, by substituting a portion of carbohydrates (CHO) by lipids (L) (keeping normal or low protein level (P) so that energy is obtained preferably from the high lipid portion, had nothing to do with MCTs in the beginning, and its objective was the treatment of people affected by convulsions with the aim of influencing neurotransmitters and the production of glutamine, which caused them, in other words, it was an anticonvulsant diet.

The aim of this diet was to maintain the $L/(CHO + P)$ ratio as high as possible, without exceeding the possible limits of fat assimilation by the body and without creating intense adverse collateral reactions.

From the point of view of its composition, and in view of the side effects it may cause, two basic types of ketogenic diets were empirically established, determined by the ratio of $L/(CHO + P) = 4/1$ in the most drastic, and 3/1 in the weakest, or most tolerable.

As can be seen, the consumption of fats was greatly exceeded, without distinction in the type of triglycerides that composed it, in fact common fats composed of long chain fatty acids were used (14-18 carbon atoms).

The history of this diet dates back to the beginning of the twentieth century, and was related to the observation that under fasting conditions, with glucose defect, patients affected by convulsions suffered less from these disorders, especially epilepsy, when this fact caused the formation in the body of a state of ketogenesis with the formation of bodies or ketonic substances.

Epilepsy is a chronic brain disease characterized by the appearance of convulsions, which are caused by excessive electrical discharges of brain cells, which can be found anywhere in the brain. Undoubtedly, these cells behave abnormally and any act or element that favours the proper functioning of them could perhaps play a positive effect for the body.

The most significant contributions in these years were those of R. Woodyatt and R. Wilder (1,2) published indistinctly in 1921, who considered that without resorting to fasting one could reach this state of ketonuria, substituting part of the carbohydrates for fats in what is equivalent to their energy contribution. Both scientists came independently to establish proportions in the use of different types of food groups and to formulate diets, the principles of which are still in force today.

According to R. Wilder's suggestions, total energy expenditure was to be achieved with 1 g of protein per kg body weight, 10-15 g of carbohydrates and the rest in fat.

In a more current sense, approximately the diet should consist of 71% fat, 19% carbohydrate and 10%

protein.

This diet continued to be used with some success for the therapeutic treatment of conditions related to seizures until the emergence of derivatives of hydantoin (phenylhydantoin) in the mid-30s of last century, which began to be used as drugs for the treatment of epilepsy and other related diseases.

Hydantoin   Difenilhydantoin

Although diphenylhydantoin (diphenylidine) was discovered in 1908 by H. Biltz (3) it was not until 1938 that H. Merryt and T. Putnam discovered that it was effective in convulsive states, so the ketogenic diet began to yield ground to this type of drugs given the unpleasant symptoms that accompanied this type of treatment, especially in its initial stage.

Nevertheless, towards 1971 P. Huttenlocher and collaborators (4) proposed the substitution of the lipids constituted by long chain triglycerides for those of medium chain, obtaining satisfactory effects, because these lipids are more easily metabolized in the organism with a greater production of ketonic bodies, with what which diet could be less drastic and more effective, and with fewer secondary disorders in terms of tolerance by people.

Earlier, in 1967 O. Owen and co-workers (5) had postulated that:

"β-Hydroxybutyrate and acetoacetate utilization by the brain shows that fat products may even satisfy the central nervous system's substrate requirement, and therefore circumvent the need for gluconeogenesis and concomitant nitrogen depletion". Aspect also considered by E. Drenick (6) in 1972.

In 1976 P. Huttenlocher himself (7) published an article in which he compared the results of the conventional ketogenic diet with those obtained when it incorporated MCT in a study carried out in humans (children) in which patients "did not show elevations of serum cholesterol and had only a slight rise in serum total fatty acids, in contrast to the marked hyperlipidemia observed in children on the standard high fat diet..."
"Long term use of the MCT diet did not affect pH of venous blood. Blood glucose fell below 50 mg/100 ml in one-third of the children, the lowest levels being reached 2--3 weeks after the start of the diet. Plasma D-beta-hydroxybutyrate (BHB) and acetoacetate rose gradually after institution of diet therapy, maximum levels being reached after about 1 month. .. Plasma BHB and acetoacetate levels in children maintained on a 3:1 high fat diet were similar to those in children on a 60% MCT diet. Plasma levels of BHB showed a significant correlation with anticonvulsant effect (p less than 0.02). Both the ketonemia and the anticonvulsant action were reversed rapidly by intravenous infusion of glucose."

Although the reader may find the references to scientific articles that we have exposed a little tedious or repetitive, as well as others that follow, it is necessary to point out that if something has been missing in the multiple controversies and disquisitions around the functional effects of coconut oil, its constituent MCTs, and its possible health benefits, it is to approach these with scientific arguments, rather than with narrations of common experiences or particular cases, so it is necessary, if not to go deeper, at least to resort to the biochemical bases of the problem and the references of scientific studies carried out by specialists on the subject, it is necessary, if not to go deeper, at least to resort to the biochemical bases of the problem and the references of scientific studies carried out by specialists on the subject.

It must be borne in mind that coconut oil is composed of more than 60% medium chain triglycerides, an issue not inherent to other fats with the exception of those of palm almonds. This undoubtedly made the scientific community pay attention to them, since from this moment on ketogenic diets would again take a prominent place in the treatment of convulsive diseases.

Thus, years later from studies by Huttenlocher and other specialists on the use and effect of MCTs in the ketogenic diet, in 1995 G. Mitchell and colleagues (8) summarized that: "Ketone bodies are produced in the liver, mainly from the oxidation of fatty acids, and are exported to peripheral tissues for use as an energy source. They are particularly important for the brain, which has no other substantial non-glucose-derived energy source. The 2 main ketone bodies are 3-hydroxybutyrate (3HB) and acetoacetate (AcAc). Biochemically, abnormalities of ketone body metabolism can present in 3 fashions: ketosis, hypoketotic hypoglycemia, and abnormalities of the 3HB/AcAc ratio. Normally, the presence of ketosis implies 2 things: that lipid energy metabolism has been activated and that the entire pathway of lipid degradation is intact".

In 2001 R. Veech in collaboration with C. Kashiwaya, H. Lardy and G. Cahill Jr. (9) published that: "...D-β-hydroxybutyrate (abbreviated "DHB") may also provide a more efficient source of energy for brain per unit oxygen, supported by the same phenomenon noted in the isolated working perfused rat heart and in sperm. It has also been shown to decrease cell death in two human neuronal cultures, one a model of Alzheimer's and the other of Parkinson's disease. These observations raise the possibility that a number of neurological disorders, genetic and acquired, might benefit by ketosis".

Later, in 2003 two of the authors of the previous article: G. Cahill Jr. and R. Veech (10) returned to the subject and stated that: "Recent studies have shown that D-beta-hydroxybutyrate, the principal "ketone", is not just a fuel, but a "superfuel" more efficiently producing ATP energy than glucose or fatty acid". They also reported that: . "Efforts are underway to prepare esters of beta-hydroxybutyrate which can be taken orally or parenterally to study its potential therapeutic applications"

Years later, in 2010 M. Samoilova and colleagues (11) reported that: "The ketogenic diet (KD), used successfully to treat a variety of epilepsy syndromes in humans and to attenuate seizures in different animal models, also provides powerful neuroprotection in various CNS injury models. Yet, a direct role for ketone bodies in limiting seizure and neuronal damage remains poorly understood. Using organotypic hippocampal slice cultures, we established an in vitro model of chronic ketosis for parallel studies of its neuroprotective and anti-convulsant effects. Chronic in vitro treatment with a ketone body, D-beta-hydroxybutyrate, protected the cultures against chronic hypoglycemia, oxygen-glucose deprivation, and NMDA-induced excitotoxicity..."

In 2003 L. Massieu and colleagues (12) noted that: "Glucose is the main substrate that fulfills energy brain demands. However, in some circumstances, such as diabetes, starvation, during the suckling period and the ketogenic diet, brain uses the ketone bodies, acetoacetate and beta-hydroxybutyrate, as energy sources. Ketone body utilization in brain depends directly on its blood concentration, which is normally very low, but increases substantially during the conditions mentioned above".

Acetoacetate effectively protects against the neurotoxicity of glutamate both *in vivo* and *in vitro*, probably through a mechanism that implies its role as an energy substrate.

Under these criteria, the ketogenic diet again began to take on relevance as a pathway for the treatment of brain disorders, primarily epilepsy, but now in an easier and more effective way to treat with the incorporation of MCTs.

Finally, in 2016 Y. Nonaka and collaborators (13) studied the action of lauric acid present in high proportion in coconut oil (about 50%), for the production of ketonic bodies in KT-5 astrocytes, which they noted occurred in much higher amounts than with oleic acid, so they considered that this medium chain fatty acid could be useful in ketogenesis.

According to these authors: "The lauric acid treatments increased the total ketone body concentration in the cell culture supernatant to a greater extent than oleic acid, suggesting that lauric acid can directly and potently activate ketogenesis in KT-5 astrocytes. These results suggest that coconut oil intake may improve brain health by directly activating ketogenesis in astrocytes and thereby by providing fuel to neighbouring neurons".

These Japanese scientists also observed that: "Initially, the ingestion of coconut oil did not substantially elevate the levels of ketone bodies in the blood, but the concentration of free fatty acids of medium chain such as lauric", so when treating the brain astrocytes of rats for four hours they found that remarkable elevation. The comparison of oils was made between coconut oil and high oleic sunflower oil because of the high concentrations of this unsaturated acid that has high oleic oil.

All of the above is based on the fact that coconut oil is attracting both media attention and potential therapy for the treatment of brain encephalic diseases, taking into account its high content of medium chain triglycerides that create ketonic bodies, which can be led to the brain and compensate for the limitations of glucose oxidation as a source of energy for neurons and thus prevent their death.

The comparison regarding the greater production of ketogenic bodies by MCT in comparison with long chain triglycerides (TCL) coming from high oleic oils and that these reached the astrocytes is very significant, since these border with the neurons to which they can provide this form of fuel of great energy efficiency, which means that coconut oil can be useful for health and good brain functioning.

In relation to the previous thing, it is necessary to consider that the astrocytes originate in the first phases of the development of the central nervous system, are directly associated with the neurons and conform the border between the organism and the cerebral system; they intertwine around the neuron and form a network of support of these, as well as act as a protective membrane of the rest of the organism and control the passage of nutrients, so the appearance of ketogenesis in them from lauric acid contained in coconut oil, is a conclusive proof of the possible action of this oil on brain encephalic functions.

Returning to the ketogenic diet, although there are several criteria, it is considered that one third of the energy content of the diet should be correlated with the MCTs in order to favour better digestive tolerance.

Medium chain triglycerides are those containing between 6-12 carbon atoms in the hydrocarbon chain, although in nature there are generally only those of even form: 6 (capric acid), 8 (caprylic acid), 10 (capric acid) and 12 (lauric acid). The latter three are the most abundant in nature, especially in almond oils from palm trees and, most of all, in coconut oil. Seen in this way, there is evidence of the positive effect of coconut oil as a source of MCT and generator of ketone bodies to prevent the conditions treated with the ketogenic diet and as a component of these as about 60% of this type of acids are found in it.

Ketone bodies, mainly β-hydroxybutyrate and acetoacetate, are the main brain fuel alternative to glucose.

Aceton

Ketoacetic acid

β-hydroxybutyric acid

The latter two are generally cited as acetoacetate and D-β-hydroxybutyric, respectively.

Ketone bodies are considered to form in the mitochondria of liver cells in several stages from acetoacetyl-CoA, which condenses with a similar molecule to produce β-hydroxy-β-methylglutaryl-CoA, which is subsequently hydrolyzed to form acetyl-CoA and acetoacetate and the latter can be reduced to β-hydroxybutyrate and also derived into acetone, although in much smaller amounts, but in general in ketogenesis the main role is played by acetoacetate and β-hydroxybutyrate.

It seems that the smaller the size of the hydrocarbon chain in MCTs, the more the ketogenic properties are accentuated and the lesser the amount of fat that can make up this diet, which led pharmaceutical chains to produce drugs containing high amounts of these triglycerides or some that are essentially only these, such as those called by their own name MCT, which are produced by fractionation of coconut oil and African palm kernel.

With what has been said so far, it would be more than conclusive to demonstrate the functional properties of coconut oil, at least for this type of affection, not counting the fact that the oil itself is sold in pharmacies, although logically protected by brand labels with prices that multiply those of supermarkets or retail product establishments.

Under any type of serious trademark, there should be no difference between coconut oil sold in pharmacies and that of retail outlets, as both should be subordinated to the "stan" of standards of the International Olive Committee (COI) and other similar institutions, as discussed in the chapter on the chemical composition of coconut oil.

Drugstores or soft capsules containing coconut oil are also sold in pharmacies under the protection of large pharmaceutical firms, which are very easy to digest, but whose oil content is relatively small in relation to the doses that can apparently be effective.

Although it is generally considered that fats provide energy equivalent to 9 kcal/g, the truth is that this depends on their composition or lipid profile, so that in MCTs this value is much lower: 7.84 kcal/g, which is a significant difference and considering that they are more ketogenic than the corresponding TCL, with much less fat can reach a state of ketogenesis, so there is less risk for obesity.

In the ketogenic diet, an amount of MCT of about 54 ml can produce an average of 420 kcal, which can be distributed in different foods to make it more tolerable to taste.

From Hunterlocher's researches, the ketogenic diet started to take relevance again, but this time accompanied by the MCTs, in this sense drugs were elaborated such as the "MCT Oils" whose constitution is generally: fractionated coconut oil, fractionated palm fruit oil, demineralised water and emulsifier E472c. This product with a high concentration of medium chain triglycerides provides 8.55 kcal/ml of lesser magnitude than the caloric contribution of conventional fats.

Other products rich in MCT are emulsions such as the "liquigen", which is a dietary preparation with high energy content consisting of MCT (capric acid (C10:0) and caprylic (C8:0)) that is said to provide 4.5 kcal / ml, by the smaller size of the hydrocarbon chains of fatty acids that compose it.

The side of formulations to make pharmaceutical preparations of MCT or products derived from same has increased significantly in recent times, so we can talk about:

Caprenin: It is nothing more than a triacylglyceride

enriched with capric acid, caprylic acid and behenic acid in its links with glycerine (Proter & Gamble) as an emulator of the properties of coconut oil, so that its activity was more intense. The content of behenic acid oscillates around 50% and its caloric contribution is between 4-5 kcal/g. The results of this product were not as expected due to its negative effect on the COLt/HDL ratio, a risk factor in cardiovascular diseases (CVD), which was discontinued at the end of the last century.

Salatrim: (short and long acyltriglyceride molecule) in this low caloric density preparation, the long chain acylglycerides (TCL) tend to be replaced by short chain acylglycerides: acetic, propionic and butyric (triacetin, tripropionin or tributyrin). The caloric intensity of these preparations is around 5 kcal/g although, the long chain fatty acid most commonly used as a complement to these joints with short-chain acids is stearic acid which is not supposed to create a risk of CVD.

Also in these preparations can include other longer chain fatty acids than the above obtaining a wide variety of forms and products, but always looking for a balance not to produce solid compounds or high caloric content, which is achieved by controlling the levels of stearic acid and short chain acids. In this way, salatrim emulates coconut butter in its physical properties for use in food products.

In our opinion, due to the high concentrations of MCTs in these preparations, it is recommended that their use requires consultation with the family doctor or other type of specialized and knowledgeable health professional. For its part, the United States Food and

Drug Administration (FDA) consider that its use should appear on food labels, while the EU approved its use in 2003. In this case, more than an MCT-producing agent, it is a low calorie food additive that provides short-chain fatty acids.

Sucrose esters have also been developed with MCTs and other fatty acids, rather related to low-calorie foods with possible incidence in the treatment of obesity such as the "olestra", an ester, or rather polyester of sucrose with fatty acids between 6-8 carbon atoms and other long chain.

In all these types of product, the FDA requires that they be labelled or require, where appropriate, authorisation from the European institutions if they are used within the territory of this community.

In relation to ketogenic products that are formed in the metabolism from MCTs, the lines have been inclined preferentially to DBH, so that it is marketed to provide an exogenous source of this ketonic body, in addition to that which the body itself produces from MCTs. The efficacy of exogenous DBH is under study, as well as other compounds derived from it.

As we have been able to analyze, there are currently different pharmaceutical formulations rich in MCTs, related to the formation of ketogenesis and in which it is a question of using the least possible amount of drugs, but with a greater effect, but always taking as a base the ketone bodies that are formed by the metabolism of medium chain fatty acids.

MCTs are easily absorbed and metabolized by the liver to produce ketones such as acetoacetate and D-β-

hydroxybutyrate, which can compete with glucose to produce energy, especially in areas such as the brain, where nerve cells are difficult to find other sources of energy, except glucose, which in certain circumstances is difficult to oxidize by insulin, such as degenerative brain diseases or diseases such as Alzheimer's disease.

It is inferred by some specialists that DBH (D-β-hydroxybutyrate) produces more energy per unit of mass than glucose, which is released in greater amounts and is something that for brain cells can be very important, since they are more exposed to these evils than other cells of the body, which can get energy by other means.

From the redox point of view, glucose: $C_6H_{12}O_6$ is a compound whose average number of oxidation for carbon is higher than in DBH ($C_4H_8O_3$), so the latter is in a more reduced state as shown in the following calculation:

Glucose:

$6C + 12H - 6O = 0$; $6C + 12 - 12 = 0$; $C = 0/6 = 0$.

Para el **DBH**:

$4C + 8H - 3O = 0$; $4C + 8 - 6 = 0$; $C = -2/4 = -0,5$.

For this type of calculation, the hydrogen oxidation number is taken to be +1, and -2 for oxygen.

In other words, ketone bodies such as DBH are a magnificent source of energy for cellular respiration and this has been shown to occur in cardiac and skeletal muscle cells in addition to the brain.

Once ingested, MCTs are absorbed by the intestine and pass into the portal circulation and directly into the liver for rapid oxidation, do not require carnitine palmitoyltransferase for mitochondrial transport, nor are they incorporated as a lipid reserve and are used immediately.

Medium-chain triglycerides offer the advantage of being more effective in preserving brain function under hypoglycemia than long-chain triglycerides. Lauric acid has been shown to increase ketone body concentrations more than other fatty acids.

REFERENCES.

(1) Woodyatt, R. (1921). *Objects and method of diet adjustment in diabetics.* Arch Intern Med 28:125–141.

(2) Wilder, R. (1921). *The effect on ketonemia on the course of epilepsy.* Mayo Clin Bull2:307.

(3) Biltz, H. (1908). *Über die Bromierung des Diphenylglyoxalons. II.* Ber. dtsch. Chem. Ges. 41, 1379 [1908].

(4) Huttenlocher, P., A. Wilbourn and J. Signore (1971*). Medium-chain triglycerides as a therapy for intractable childhood epilepsy.* Neurology.1971 Nov; 21(11):1097-103.

(5) Owen, O, et al. (1967). *Brain metabolism during fasting.* J Clin Invest 1967; 46:1589–95.

(6) Drenick, E, et al. (1972). *Resistance to symptomatic insulin reactions after fasting.* J Clin Invest 1972; 51:2757–62.

(7) Huttenlocher, P. (1976). *Ketonemia and seizures: metabolic and anticonvulsant effects of two ketogenic diets in childhood epilepsy.* Pediatr Res. 1976 May;10(5):536-40.

(8) Mitchell, G. et al. (1995). *Medical aspects of ketone body metabolism.* Clin Invest. Med.1995 Jun; 18(3):193-216.

(9) Veech, R. et al. (2001). *Ketone bodies, potential therapeutic uses*. IUBMB Life. 2001. Apr; 51(4):241-7.

(10) Cahill G.Jr. and R. Veech (2003*). Ketoacids? Good medicine*? Trans Am Clin Climatol Assoc. 2003;114:149-61; discusión 162-3.

(11) Samoilova M1, et al. (2010). *Chronic in vitro ketosis is neuroprotective but not anti-convulsant.* J.Neurochem. 2010 May; 113(4):826-35.

(12) Massieu, L. (2003). *Acetoacetate protects hippocampal neurons against glutamate-mediated neuronal damage during glycolysis inhibition.* Neurosciense. 2003; 120(2):365-78.

(13) Nonaka, Y. et al. (2016). *Lauric Acid Stimulates Ketone Body Production in the KT-5 Astrocyte Cell Line.* (J Oleo Sci. 2016 Aug 1;65(8):693-9.

CHAPTER IV

COCONUT OIL AND OBESITY

It can be inferred that after finding that the medium chain fatty acids found in coconut oil do not follow the same metabolic route in the body as the other long chain fatty acids, to which is added that their oxidation is faster and they produce less energy per unit of mass ingested, they should have some effect on overweight and obesity, and a last argument, perhaps the most striking: they do not form adipocytes, and therefore are not stored in adipose tissue, at least under moderate intake, at least not stored in adipose tissue.

Yes, with these arguments it is advisable to focus on

the possible use of coconut oil in a diet to lose weight, as is the case with ketogenic, but for the moment we do not consider that the latter should be the most indicated, at least until exhausting other routes, because it requires substantial reductions of other nutrients and has relatively more intense effects than a normal diet, in addition to its use is very well intended for other disorders, such as brain encephalic.

In addition to the above, there is some evidence from recent studies that coconut oil does not contribute to atherosclerotic damage, regardless of whether Col_T and LDL values are increased, since it does so to a greater extent with HDLs, thus counteracting this effect. This is very important to bear in mind, because generally with obesity are associated cardiovascular disorders.

However, dictating a coconut oil diet against overweight and obesity has the same drawback as the hundreds of diets that are proposed and recommended daily in written and oral form, and there are many, and every day some more are published, but if only one of them had the desired effect without leading the body to a state of stress or a superhuman effort, or completely alter the way of life of people, or what is worse, that can cause other disorders, then not so many diets would be needed. And it is that sometimes it is forgotten that obesity and overweight are a disorder associated with the form and habits of social life of our time.

Yes, in previous times levels of obesity and overweight were not as high as they are today, which is a concern for those who care about the health status of people and for themselves. The percentage of obese or overweight people is high, especially in developed and industrialized countries, where all the factors come together for people to gain weight and there is an imbalance between the energy consumed and that spent, falling then into an essentially thermodynamic problem, according to its first principle or "Law of conservation and transformation of energy", so that the human organism, as protection does the most advisable thing, store it in the material form that can best be according to its content per unit of mass, that is to say, the fats in the adipose tissue.

And once the fats have been stored and the imbalance has been maintained, more energy comes in than is spent on the different activities in which men participate, most of them sedentary and lacking in intense physical activity, then this apparently unnecessary hoarding continues until what is to be expected happens, this bubble burst, but not in the form in which fat disperses and disappears, but through other evils: hypertension, cardiovascular disorders, etc. some of which can be fatal.

Added to everything else is the high availability of food of very varied quality and pleasant tastes to which people can have access according to their purchasing power, especially in developed countries. And this is something that did not happen in the past.

In economically poor and underdeveloped countries, most of the purchasing power of citizens is used for food, and people have fewer means of mechanical transport, make greater journeys walking, perform work with greater physical activity, and generally in larger spaces, many of them outdoors, so they consume a greater amount of energy and in greater proportion reach thermodynamic balance.

Undoubtedly, development has its prices in addition the common diets of people in developed countries are richer in fats and proteins, and less in vegetables, thereby increasing exposure to overweight and obesity.

Added to the above, the increased consumption of processed and semi-finished products to make more bearable the pace of life, or forced by the dynamics or remoteness of work, decreasing the frequency of healthy meals and traditional care for their preparation. In a house it is cooked, not only in dependence of the tastes, but also in agreement with the well-being of the people who constitute the family nucleus; in a restaurant or in a fast food establishment the parameters are other, but in no case in relation to the state of well-being or the health of the consumer. The same thing will be served a dish of French fries with an over used oil to the thinner person than to the obese, without any distinction or warning, and cooked with the right oil or the wrong one. For the establishment this is a consumer problem.

In view of the above, and that dietary campaigns encouragement of physical exercise and others, in addition to unfounded dietary promises, according to statistics has not solved the problem, it is getting worse every day.

It is enough to have an idea about the incidence of obesity and overweight in the modern world to see that the ten countries with the highest index in the world surpass 25% of the population among people over 15 years, being the one with the most unfavourable indicator United States with 38% followed by Mexico, 32.4% and New Zealand with 30.7%. In other words, in the most developed industrial state in the world 2 out of 5 people are affected by overweight, and in the case of Mexico it can be seen that development charges a high price in the state of people's health. Something similar happens with China where between 2006 and 2016 the levels of overweight in children doubled, at the same time that industrialization and economic development were growing at an advanced pace. Something similar happens with Chile, it is a country that in recent times has made great progress in its agricultural-industrial development, but has also paid its toll on this tortuous road, and ranks 8th worldwide in this negative indicator (1).

Undoubtedly, it would be absurd to state that development must be stopped, because it brings progress and other factors of social welfare and health for the population, but it is necessary to establish that not only with diet can overweight and obesity be solved in the midst of such an aggressive climate.

There are those who simplify the problem and argue that overweight, when it is not a genetic disease, is combated with only two things: diet and physical activity, to which I would add another: the environment, and as long as it acts with such force and intensity, we doubt that the first two can be done effectively and much less so that it is what the moment demands.

The environment is also the cause of intense propagandas of food products, some of which pleasant taste and presence are a temptation for people. In the field of vegetable oils, we have noticed that propaganda is over dimensioned in this sense, and rather than suggesting that only what the organism needs is consumed, it is advisable to consume more and for anything; and coconut oil, although it is not the main protagonist in the complex olive sector, is not exempt from this; and rather than assessing which oils can be substituted, or how to balance them with others, there is a tendency towards supplementation, a matter that favours even more non-compliance with the principles of thermodynamics, which in this case must be inviolable.

It seems that thermodynamics does not only affect the First Principle, but also the Second, Entropy as a measure of disorder, and if it was postulated that systems tend to the maximum of entropy and disorder, the latter does not only refer to molecules; our world seems to be more chaotic and disordered every day, and food also follows this pattern. When a disorder has occurred in the balance and the alimentary balance it is very difficult to put order and that the things return to their place, and this they know it very well those who affirm that to gain weight is made easy, of spontaneous way, but to lose, it is very difficult.

In the case of people with normal weights, it is advisable to ingest the energy content equivalent to that which is going to be spent, and in those who are overweight, as less, equal, but preferably less to reduce this parameter.

It is convenient, before continuing, to stop in some concepts, and more than all, in the one of "obesity", that of simple form can be defined as a chronic disease, but treatable, that comes accompanied by an excess of accumulation of fats in the adipose tissue. It is the most common disorder in society in developed countries. It increases with age and its causes can be: environmental and social factors, excess food intake, especially in caloric foods, as well as genetic factors of metabolism and hormonal. It is characterized by overweight. It is one of the main cardiovascular risk factors.

In addition to weight, the easiest way to measure obesity is waist diameter: more than 35 inches for women and 40 inches for men to be considered overweight, as well as body mass index (BMI), which is calculated by dividing kilograms of weight by the square of height (BMI = P/h^2).

It is considered that the best way to treat obesity is through a change in lifestyle, with all the factors that this may entail, including physical activity and diet, in which you should consume less energy (calories) than those that are eliminated, to acquire normal weight, after which a balance must be maintained between expenditure and consumption.

If the calories consumed with the intake of food are not spent, obesity increases, if you spend the same amount that is consumed, weight is maintained, if you spend more than you consume, weight decreases.

Undoubtedly, physical exercise is the most effective way to spend energy, but as overweight increases there are more mobility difficulties and other symptoms arise that may limit its implementation.

Fats are the nutrients that contain more energy per unit of mass due to structural and thermodynamic factors, at the same time they are the most rational and less voluminous way of storing energy through adipose tissue. If energy were stored in the form of carbohydrates or proteins, the volume of people would at least double, as they have a caloric power less than half that of lipids.

To the nutritional properties of fats must be added that constitute a form of defence of the body against certain contingencies, because otherwise its excess would be evacuated during the digestive process, it is as if by an automatic mechanism the body anticipated times of lack of food or famine, as happened in the past.

Until now, the general rule established that ingesting saturated fats increased levels of overweight more than unsaturated, but now evidence is emerging that this cannot be seen as a principle without exceptions, as some recent studies and research guarantee that this is not met for all saturated fats, as those with high levels of medium chain triglycerides (MCT) do not behave in this way, as is the case of coconut oil.

Therefore, the physicochemical properties of medium and short chain saturated fatty acids, which determine how they are metabolized in the body, do not allow them to raise weight levels and be stored as adipose tissue, although if consumption is very high, this can be done, especially with the higher hydrocarbon chain as lauric. But under normal conditions, medium chain saturated fatty acids compete with other long chain fatty acids and these are the ones that come to be stored in adipose tissue regardless of their character, although preferably the most frequent saturated in the diet: palmitic (C16:0), stearic C18:0) and the lesser of them; the myristic (C14:0).

Therefore, it can be inferred that ingesting coconut oil, rather to introduce it in the meals as a more oil, could result in improvements for the overweight given in addition to its tendency not to store its medium chain fatty acids, in addition to the fact that they have less energy power, among other factors, so that per unit of mass evolve fewer calories. However, adding coconut oil as a supplement if the same calorie content is maintained with other fats, will not result in a reduction in weight, as the latter will be stored and not consumed, can even have an adverse effect on raising the consumption of calorie energy, so the ideal is the partial replacement (better than total) of other oils, we prefer the partial maintaining adequate levels of high oleic fats (olive oil, high oleic sunflower, among others) to protect the CVD, the myristic acid and other saturated fatty acids contained in coconut oil, and the need for the diet to be varied and provide as much nutrients as possible, including essential polyunsaturated fatty acids: linoleic and linolenic acids that are not synthesized by the organism.

From the electrochemical point of view, the long hydrocarbon chains of the fatty acids are more reduced so they tend to produce more energy by oxidation than those of medium and short chain acids.

For example, for palmitic acid:

Palmitic acid: $CH_3(CH_2)_{14}COOH$: $C_{16}H_{32}O_2$, the average number of oxidation for the carbon in the molecule is:

$16C + 32 - 4 = 0$, $C = -(28/16) = -1.75$

While for **caprylic acid**: $CH_3(CH_2)_6COOH$: $C_8H_{10}O_2$, the average oxidation number for the carbon in the molecule is:

$8C + 14 - 4 = -(10/6)$; $C = -1,66$

According to these values, a long chain saturated fatty acid such as palmitic acid is: $1.75/-1.66 = 1.05$ times less than a medium chain reducer such as caprylic.

Also, in addition to redox factors and those related to lipogenesis, it should not be forgotten that MCTs, especially smaller ones, are a source that favours the formation of ketone bodies. This mechanism of producing ketonic bodies is less given for long chain fatty acids, so that the former produce about four times more than the latter and their energy expenditure, if consumed moderately, is immediate and in the following hours of being ingested.

This is related to the fact that in the metabolism of MCTs, they are transported directly to the liver where they are oxidized to ketone bodies: acetoacetate (CH_3COCH_2COO) and D- β-hydroxybutyrate ($CH_3CHOHCH_2COO$), while those of longer length can form acetyl-CoA and pass to the respiratory chain and the Krebs cycle. Ketone bodies can go to other tissues, including the brain, to produce energy, a process that, due to its smaller size and complexity, develops faster.

On the other hand, ketone bodies show a certain tendency to create satiety, which enables a better control of the diet for people with overweight disorders.

Notwithstanding the above, the calculations lead to the need to consume high proportions of AGSCM to obtain significant weight reductions, with complex results, since they can act on other parameters such as triglycerides (TG) and cholesterol and its response to treatment.

Despite the latter, the effect of MSCA on obesity and overweight and associated factors has been carefully observed for some years now; and since coconut oil is a natural product containing high concentrations of these acids, it is receiving attention from the international community, and in particular from the scientific community (2).

It is therefore necessary to evaluate the results of research carried out in recent years in this field, in order to draw the relevant conclusions, and what better than to begin by referring to the results published in 1993 on a famous population-based follow-up study, related to the particular habits and lifestyle of the inhabitants of the island of Kitava, the archipelago of Trobriand Island in the Pacific Ocean (Papua New Guinea) (3).

The inhabitants of this island, who maintained (and we say maintained because the current situation should not be the same) a primitive lifestyle of subsistence, based their diet on fruits, vegetables, including coconut, and fish. So these two latter were their main source of fat. There the study was carried out with all the population (1816 people) and between the measured parameters and the observations and reports of its inhabitants it was concluded that: "The vascular brain accident and the ischemic cardiopathy seem to be absent in this population", and also other evils related to the modern lifestyle.

On the other hand, different studies have been carried out in animals, mainly in rats, on the relationship between obesity and MCTs, including the one published in 1980 by G. Bray and collaborators (4). In it, they compared weight gain in rats fed corn oil (rich in polyunsaturated fatty acids) and MCT to assess the metabolic route of both, and the effects they could cause, since in one circulate as chylomicrons and in the other go directly to the liver by portal circulation, finding that there was a greater weight gain with that of diets rich in corn oil than with those high in MCT. Also the caloric intake was higher in this one. In summary, they "…suggest that the route by which nutrients are absorbed plays a role in regulating body-fat storage".

Later, to determine the effect of an overfeeding of MCTs in rats compared to diets rich in long chain triacylglycerides (LCTs), A. Geliebter and colleagues (5) published in 1983 the results of an experiment in which both types of fats were administered to cover 45% of the total energy expenditure. The results obtained were as expected: rats fed MCTs gained 20% less weight and showed fat deposits that weighed less than 23%; also the average size of the adipocytes was smaller than in those using long chain fatty acids, implying that MCTs might be suitable for reducing obesity in humans.

In 1987 the results of very complete experiments of the parameters associated with MCT and LCT diets in the presence or not of carbohydrates were published (6). It was found that the rats fed with higher MCT content had an increase of less than 30% in weight than the others, as well as energy retention, which led to a 60% decrease in daily lipids. Serum concentrations of ketone bodies were higher in the MCT-enriched diet, but decreased throughout the experiment until the end to be half of that obtained in the initial stage, which could be due to an adaptation of the rats to a diet rich in MCT. During the study, other parameters of interest were measured, such as DBH/acetoacetate ratio, lactate/pyruvate, malic enzyme activity in the liver, among others.

The structural character of fats in medium- and long-chain triglycerides and their effect on body fat in rats were measured in comparison with normal long-chain triacylglycerides (7), obtaining that the fat content of intra-abdominal tissues and the carcass were lower than in rats fed a diet rich in the latter, so it is concluded: that the former are less effective for fat accumulation in adipose tissue and therefore tend not to favour obesity.

To determine the differentiated effect of thermo genesis between MCT and LCT, in 2002 Japanese scientists (8) studied the calculation of oxygen consumption by rats fed both types of oils and their results - of great interest - showed that this was higher in rats fed MCT, as well as that the abdominal fat content was lower than in those fed LCT. Measured in calorie terms MCTs decreased by 0.27 kcal/g of fat more than TBIs.

On a human level, there are references from 2009 to a double-blind study with 40 women between 20 and 40 years of age characterised by abdominal obesity, who were given 30 ml of coconut or soy oil, indistinctly, for twelve weeks, under a balanced hypo caloric diet accompanied by certain physical activity (9). The results obtained showed that in the group with coconut oil the LDL increased more than in the soy group (Very rich in PUFA), so a lower LDL/HDL ratio was obtained; and although there were decreases in the abdominal diameter in both groups, in the coconut oil the latter was more significant, which led to the conclusion that coconut oil did not produce hyperlipidemia and contributed to a decrease in abdominal obesity.

An interesting anthropometric study that relates the effect of different vegetable oils on obesity was recently carried out in Brazil (10), in which the action of coconut, chia, safflower and soybean oils on this disorder was compared; and it was found that: "Coconut oil had a more pronounced effect on abdominal adiposity and glycidic profile, whereas chia oil had a higher effect on improving the lipid profile. Indeed, supplementation with different fatty acid compositions resulted in specific responses".

An interesting comparison of the effects of diets rich in MCTs vs. others with LCT was carried out in 2003 to compare the effect of both during a 27-day study in overweight women (11). The diets in question contained 40% energy in the form of fats; the MCT in the form of a preparation of very similar parts of octanoate (capric) and decanoate (caprylic), and the LCT as beef tallow. Body composition measurements were made using nuclear magnetic resonance techniques. In the end, it was found that long-term consumption of MCT improved ES (energy efficiency) and fat oxidation in obese women under study, compared to consumption of TBI. The difference in body composition change between MCT and LCT consumption, although not statistically significant, was consistent with the differences predicted by the changes in the US. It can be concluded that substituting MCT for LCT in an energy-balanced diet may prevent long-term weight gain through a higher EE.

In 2016, a randomized meta-analysis was conducted on trials conducted to measure the comparative effects of LTCs and MCTs on overweight indicators, finding that: "Replacement of LCTs with MCTs in the diet could potentially induce modest reductions in body weight and composition without adversely affecting lipid profiles. However, further research is required by independent research groups using large, well-designed studies to confirm the efficacy of MCT and to determine the dosage needed for the management of a healthy body weight and composition". (12). No significant differences were found for plasma lipids.

In 2010, the effect of fatty acid chain length, postprandial and food intake in thin men was investigated to see if medium chain triglycerides (MCTs) showed a greater decrease in appetite, given their greater kinetics of oxidation accompanied by attenuated lipemia. Coconut oil was used as a diet with a high MCT content, and tallow was used as fat for LTCs. The measured parameters, including perceived pleasantness, visual appearance, smell, taste, taste and palatability, did not show significant differences between the two groups, so; it is concluded "...that was no evidence that fatty acid chain length has an effect on measures of appetite and food intake when assessed following a single high-fat test meal in lean participants". (13)

With regard to satiety, which is a parameter that can play an important role in food intake, comparative trials were made between oils rich in MCT, with conjugated linoleic acid (C18:2) in about twenty healthy adults. The time between meals was measured as well as the satiety of the visual analogue scales; and the results showed that both oils produced this effect in different foods, without significant differences between the two, which increased satiety and decreased energy intake (14).

As the use of coconut oil for different functional purposes has become relevant in recent years and these have a high concentration of MCTs, it could be interesting to compare its effect with that of fats very rich in MCTs in increasing satiety, which translates into lower food intake in practice, and consequently, reducing obesity. For this purpose, human trials were carried out, the results of which indicate that MCTs increased significantly more obesity than coconut oil, with which people consumed less food; although coconut oil showed better results than the control group. (15) In this way, it is verified that the substitution of the MCT by coconut oil to cause satiety and decrease the food intake is not completely feasible, although this shows incidence on satiety. It is not really comparative its use in this sense, considering that in preparations rich in MCT there are more of them and generally shorter chain length (C6:0, C8:0, etc.) not as in coconut oil where lauric acid predominates (C12:0) of longer hydrocarbon chain length.

Given the results of previous research on the effect of coconut oil and MCTs on obesity and its associated factors, there is sufficient evidence that they have a favourable impact on weight loss, but the level of their contribution does not seem to be in a practical way sufficient to achieve adequate weight levels for overweight people, at least immediately. However, they do contribute to a certain extent to their reduction, and MCT preparations are, as expected, much more effective than coconut oil. This does not exclude the use of this oil by people as long as they replace other long chain fats in the same amount of energy contribution of the same, and with levels according to the possibilities of the body, so it would be advisable to follow the advice of your doctor.

Coconut oil and MCTs alone cannot solve the problem of overweight and obesity, because of the multiplicity of factors that influence them, are not a miracle as would have been desired, although they are better than their long chain congeners which are those commonly consumed by the population, and contribute somewhat to reducing this uncomfortable and harmful affection that increasingly affects a high portion of human society in developed countries. For all the above reasons, its inclusion as a cooking and salad oil seems more than advisable, although preferably conjugated with other oils with a high content of unsaturated fatty acids.

REFERENCES:

(1) El Comercio, Perú. (2018). *¿Cuál es el país con mayor índice de obesidad?* Redacción EC 08.04.2018.

(2) Sayazo-Ayerdi, S., et al. (2008). *Utilidad y controversias del consumo de ácidos grasos de cadena media sobre el metabolismo lipoproteico y obesidad.* Nutr. Hosp. 2008;23(3):191-202

(3) Lindeberg. S. and B. Lundh. (1993). *Apparent absence of stroke and ischaemic heart disease in a traditional Melanesian island: a clinical study in Kitava.* J Intern Med. 1993 Mar; 233(3):269-75.

(4) Bray, G, M. Lee and T. Bray. (1980). *Weight gain of rats fed medium-chain triglycerides is less than rats fed long-chain triglycerides.* Int J Obes. 1980;4(1):27-32.

(5) Geliebter A., et al. (1983). *Overfeeding with medium-chain triglyceride diet results in diminished deposition of fat.* American Journal of Clinical Nutrition 37(1):1- 4. February 1983.

(6) Gayle Crozier, et al. (1987). *Metabolic effects induced by long-term feeding of medium-chain triglycerides in the rat.* Metabolism Volume 36, Issue 8, August 1987, Pages 807-814

(7) Tatsuhiro Matsuo and Hiroyuki Takeuchi (2004). *Effects of structured medium- and long-chain triacylglycerols in diets with various levels of fat on body fat accumulation in rats.* British journal of nutrition Volume 91, Issue 2 February 2004 , pp. 219-

(8) Osamu Nogushi, et al. (2002). *Diet-Induced Thermogenesis and Less Body Fat Accumulation in Rats Fed Medium-Chain Triacylglycerols than in Those Fed Long-Chain* Triacylglycerols. J Nutr Sci Vitaminol, 48, 524-529, 2002.

(9) Assunção M. et al. (2009) *Effects of dietary coconut oil on the biochemical and anthropometric profiles of women presenting abdominal obesity. Lipids.* 2009 Jul; 44(7):593-601.

(10) Oliveira-de-Lira L. et al. (2018). *Supplementation-Dependent Effects of Vegetable Oils with Varying Fatty Acid Compositions on Anthropometric and Biochemical Parameters in Obese Women.* Nutrients. 2018 Jul 20;10(7). pii: E932.

(11) St-Onge M., et al. (2003). *Medium- versus long-chain triglycerides for 27 days increases fat oxidation and energy expenditure without resulting in changes in body composition in overweight women.* Int J Obes Relat Metab Disord. 2003 Jan; 27(1):95-102.

(12) Mummet, K. and W Stonehouse. (2015). *Effects of Medium-Chain Triglycerides on Weight Loss and Body Composition: A Meta-Analysis of Randomized Controlled Trials.* Journal of the Academy of Nutrition and Dietetics. Volume 115, Issue 2 February 2015, Pages 249-263.

(13) Poppittab S., et al. (2010). Fatty acid chain length, postprandial satiety and food intake in lean

men. Physiol Behav. 2010 Aug 4;101(1):161-7.

(14) Coleman H, P. Quinn and M. Clegg. (2016). *Medium-chain triglycerides and conjugated linoleic acids in beverage form increase satiety and reduce food intake in humans.* Nutr Res. 2016 Jun;36(6):526-33.

(15) Kinsella R., T. Maher and M.Clegg. (2017) *Coconut oil has less satiating properties than medium chain triglyceride oil.* Physiology & Behavior Volume 179, 1 October 2017, Pages 422-426

OTHER PUBLICATIONS OF THE AUTHOR

ACEITE DE COCO

CALIXTO LÓPEZ HERNÁNDEZ

ACEITE DE AGUACATE

CALIXTO LÓPEZ HERNÁNDEZ

CALIXTO LÓPEZ
ROSALÍA ROJCO

ACEITE DE ALGODÓN

OTHER BIBLIOGRAPHIC SOURCES

Adkins, ed. S.; M. Foale and Y. Samosir (2006). *Coconut revival new possibilities for the 'tree of life'*: Proceedings of the International Coconut Forum held in Cairns, Australia, November 2005.

AOCS. (1997). Official *Methods and Recommended Practices of the American Oil Chemists Society, 5th ed.* D. Firestone (ed), AOCS Press, Champaign.

Astiasarán, Y. y J. Martínez, (2003). *Alimentos. Composición y propiedades.* McGraw-Hill Interamericana. Madrid.

Bach, A., Y. Ingenbleek and A. Frey, (1996). *The usefulness of dietary medium-chain triglycerides in body weight control: fact or fancy?* J Lipid Res 1996; 37: 708-726.

Babayan, V. (1987). *Medium chain triglycerides and structured lipids.* Lipids 1987; 22: 417-420.

Badui, S. (2006). *Química de los Alimentos. 4ta. Edic.* PEARSON. Adison Wesley. México.

Bailey, A. (1961). *Química de los Alimentos. 3ra. Edic. Editorial.* Addison Wesley Longman. México.

Coultate, T. (1998). *Manual de Química y Bioquímica de los alimentos.* Ed Acribia. España.

Departamento de Salud y Servicios Sociales de los Estados Unidos (2010). *Dietary Guidelines for Americans.*

Drenick E, et al. (1972). *Resistance to symptomatic insulin reactions after fasting.* J Clin Invest 1972;51:2757–62.

Eldridge, J., D. Cooper and J. Peters.(2002). *A role for olestra in body weight management.* Obes Rev 2002; 3: 17-25.

Finley, J. et al. (1994). *Caloric availability of SALATRIM in rats and humans.* J Agric Food Chem 1994; 42: 495-499.

FDA (1996). *Food additives permitted for direct addition to food for human consumption; olestra, final rule.* Federal Register, Part III, 21 CFR part 172. US Department of Health and Human Services: Food and Drug Administration 1996; 61: 3118-3173.

Foale, M. (2003*). The Coconut Odyssey: The Bounteous Possibilities of the Tree of Life Canberra*: Australian Centre for International Agricultural Research. pp. 115-116.

Foster, R.; C. Williamson, and J. Lunn, (2009). *Culinary oils and their health effects.* Nutrition Bulletin 34 (1): 4-47.

Gunstone, F. (2002). *Vegetable oils in food technology.* Editor R. Hamilton. Blackwell Publishing CRC

Grimwood, B. (1979). *Coconut palm products: their processing in developing countries.* 2da. edición. Roma: FAO. pp. 193-210.

Hashim, S., A. Arteaga and T. Van Itallie. (1960). *Effect of a saturated medium-chain triglyceride on serumlipids in man.* Lancet 1960; 1: 1105-1108.

Holt, P. (1967). Medium chain triglycerides. A useful adjunct in nutritional therapy. Gastroenterology 1967; 53: 961-966.

Hu, F., et al. (1997). *Dietary fat intake and risk of coronary heart disease in women.* N Engl J Med 1997; 337: 1491-99.

Hu, F. et al. (1999). *Dietary saturated fats and their food sources in relation to the risk of coronary heart disease in women.* Am. J. Clin Nutr 1999; 70: 1001-8.

Kritchevsky D. (1998). *History of recommendations to the public about dietary fat.* J. Nutr 1998; 128: 449-52.

Kromhout D, et al. (1995). *Dietary saturated and trans fatty acids and cholesterol and 25-year mortality from coronary heart disease: the Seven Countries Study* Prev Med 24: 308-15.

Krotkiewski, M. (2001). *Value of VLCD supplementation with medium chain triglycerides.* Int J Obes Relat Metab Disord 2001; 25: 1393-1400.

Lambruschini, N., A. Gutiérrez (Coord.) et al. (2012). Dieta Cetogénica. Spanish Publishers Associates © 2012.

López, C. (2018). *Química de los Aceites Vegetales.* Amazon Kindle KDP Publishing. ISBN. 9781980870401. Spain.

López, C. (2018). *Aceite de Coco.* Amazon Kindle KDP Publishing. ISBN 978198 2999483. Spain.

Lichtenstein A, et al. (1999). *Effects of different forms of dietary hydrogenated fats on serum lipoprotein cholesterol levels.* N Engl J Med (1999); 340: 1933-40.

Moreiras O. et al. (2007). *Tablas de composición de alimentos. 11ª edición.* Pirámide. Madrid.

Mozaffarian D, R. Clarke (2009). *Quantitative effects on cardiovascular risk factors and coronary heart disease risk of replacing partially hydrogenated vegetable oils with other fats and oils.* Eur J Clin Nutr 2009; 63: S22-S33.

Oliver A. et al. (2008). *El libro blanco de las grasas en la alimentación funcional.* 2008 Unilever España, S.A. ISBN: 978-84-612-7466-6, España

Pehowich DJ, A. Gomes and J. Barnes (2000). *Fatty acid composition and possible health effects of coconut constituents.* West Indian Med J. 49,128-33.

Petrauskaité V, W. De Grey and M. Kellens (2000). *Physical refining of coconut oil: Effect of crude oil quality and deodorization conditions on neutral oil loss.* J. Am.Chem. Oil Soc. 77, 582-586.

Ranhotra, G., J. Gelroth, and B. Glaser (1994). *Usable*

energy value of a synthetic fat (caprenin) in muffins fed to rats. Cereal Chem 1994; 71: 159-161.

Rao R. and B. Lokesh (2003). *TG containing stearic acid, synthesized from coconut oil, exhibit lipidemic effects in rats similar to those of cocoa butter*, Lipids, 38, 913-918.

Siri-Tarino P., et al. (2010). *Meta-analysis of prospective cohort studies evaluating the association of saturated fat with cardiovascular disease.* Am J Clin Nutr 2010; 91: 535-46.

Stafstrom, C. and J. Rho. (2012). *The ketogenic diet as a treatment paradigm for diverse neurological disorders*. Pharmacol., 09 April 2012.

Swift, L., et al. (1992). *Plasma lipids and lipoproteins during 6 d of maintenance feeding with long-chain, medium-chain, and mixed-chain triglycerides*. Am J Clin Nutr 1992; 56: 881-886.

Taha A, S. Henderson S and W. Burnham. (2009*). Dietary enrichment with medium chain triglycerides (AC-1203) elevates polyunsaturated fatty acids in the parietal cortex of aged dogs: decline*.Neurochem Res. 2009 Sep;34(9):1619-25. Epub 2009 Mar 20.

Torrejón, C. y R. Uauy. (2011). *Calidad de grasa, arterioesclerosis y enfermedad coronaria: efectos de los ácidos grasos saturados y ácidos grasos trans*. Rev Med Chile 2011; 139: 924-931.

Warner K, and N. Michael-Eskin (1995). *Methods to asses quality and stability of oils and fat-containing foods*. AOCS Press. Illinois, USA. Cap. 2,9.

Zschau W. (2000). *Introduction to Fats and Oils Technology*, 2nd edn. Champaign, IL: AOCS Press.

ÍNDEX

Author's foreword------------------------------ Page 03

Chapter I. Introduction ------------------------ Page 05

Chapter II. Composition of coconut oil ------ Page 12

Chapter III. Coconut oil and ketogenic diet --Page 36

Chapter IV. Coconut Oil and Obesity--------- Page 55

Other Publications for the Author ------------- Page 75

Other bibliographical sources------------------- Page 82

Index -- Page 88

www.ingramcontent.com/pod-product-compliance
Lightning Source LLC
Chambersburg PA
CBHW071418220526
45469CB00004B/1321